Math Fun with Puppies and Kittens

Learning
ADDITION

with Puppies and Kittens

Enslow Publishing
101 W. 23rd Street
Suite 240
New York, NY 10011
USA

enslow.com

Eustacia Moldovo
and Patricia J. Murphy

Words to Know

addend A number added to another.

basic Simplest.

equal Of the same value.

operation A way to get one number using other numbers by following special rules.

sum The total from adding numbers.

value The amount or worth of something.

Contents

Addition

Addition is the kind of math you use to put together two or more numbers. The answer is called the **sum**. The sum tells you how many things you have all together. You can count puppies and kittens to learn how addition works!

4 + 3 is called an addition fact. There are two ways to write addition facts:

addends

$$\begin{array}{r} 4 \\ + \ 3 \\ \hline 7 \end{array}$$

$$4 + 3 = 7$$

sum

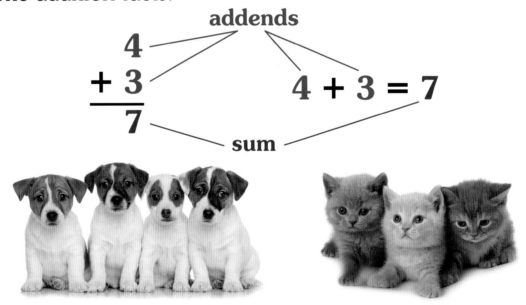

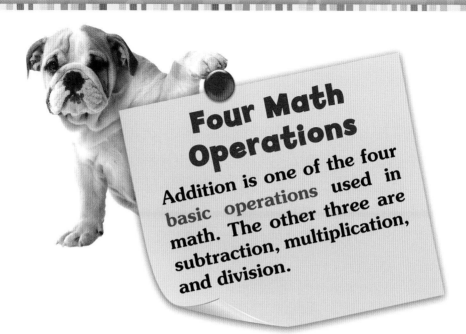

Four Math Operations

Addition is one of the four basic operations used in math. The other three are subtraction, multiplication, and division.

There are four parts to an addition fact:

(1) There are two **addends**: **4** and **3**. These are the numbers you add together.

(2) There is a plus sign (**+**). It means add.

(3) There is an **equal** sign (**=** or **___**).

(4) There is the sum after the equal sign: **7**.

Learn to Add 0

Let's start with the number 0. If you add 0 to any number, the answer will always equal that number. Nothing changes because 0 has no **value**.

2 + 0 puppies take a nap

Begin at **2**. Move **0** spaces to the right. The sum is **2**.

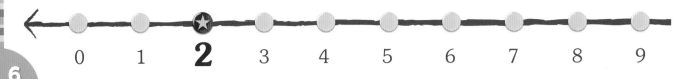

Number Lines

Number lines make adding easier. Start at the first addend. Then move right the number of spaces equal to the second addend. You will land on your sum.

$$0 + 0 = 0$$
$$1 + 0 = 1$$
$$\mathbf{2 + 0 = 2}$$
$$3 + 0 = 3$$
$$4 + 0 = 4$$
$$5 + 0 = 5$$
$$6 + 0 = 6$$
$$7 + 0 = 7$$
$$8 + 0 = 8$$
$$9 + 0 = 9$$
$$10 + 0 = 10$$

$= 2$

10 11 12 13 14 15 16 17 18 19 20

Learn to Add 1

When you add 1 to any number, the answer will always be one more than that number.

4 + 1 kittens play with yarn.

Begin at **4**. Move **1** space to the right. The sum is **5**. The number **5** is **1** more than **4**.

= 5

$$0 + 1 = 1$$
$$1 + 1 = 2$$
$$2 + 1 = 3$$
$$3 + 1 = 4$$
$$\mathbf{4 + 1 = 5}$$
$$5 + 1 = 6$$
$$6 + 1 = 7$$
$$7 + 1 = 8$$
$$8 + 1 = 9$$
$$9 + 1 = 10$$
$$10 + 1 = 11$$

10　11　12　13　14　15　16　17　18　19　20

Learn to Add 2

Counting on is one way to find the sum when you are adding more than **1**. Count on with things or a number line. To solve **3 + 2**, count on with puppies. Begin with **3**. Then "count on" **2** more.

3 + 2 puppies chew on bones.

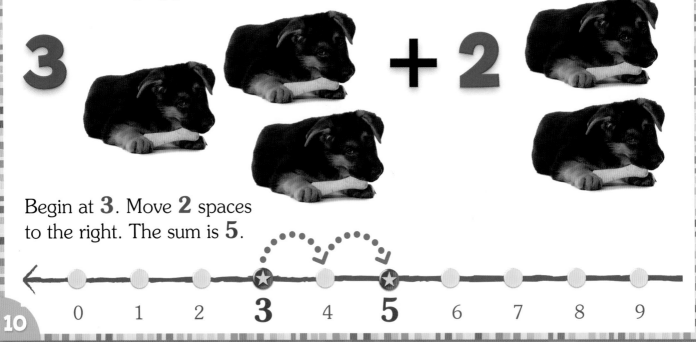

Begin at **3**. Move **2** spaces to the right. The sum is **5**.

= 5

$$0 + 2 = 2$$
$$1 + 2 = 3$$
$$2 + 2 = 4$$
$$\mathbf{3 + 2 = 5}$$
$$4 + 2 = 6$$
$$5 + 2 = 7$$
$$6 + 2 = 8$$
$$7 + 2 = 9$$
$$8 + 2 = 10$$
$$9 + 2 = 11$$
$$10 + 2 = 12$$

10   11   12   13   14   15   16   17   18   19   20

Learn to Add 3

The addition fact **3 + 3** is called a doubles fact. You add the number to itself. The sum is twice the number.

3 + 3 kittens wash their faces.

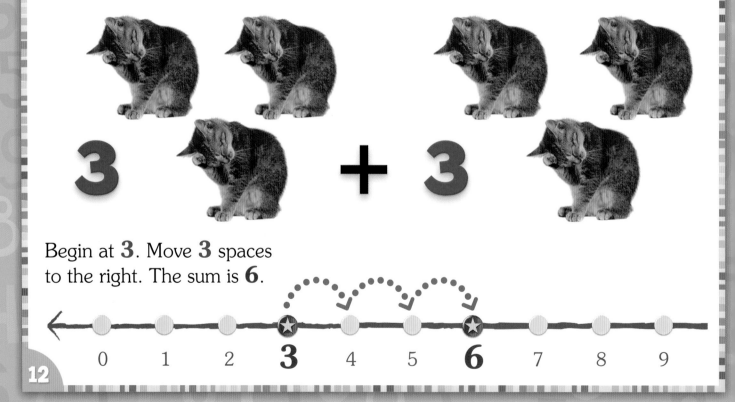

Begin at **3**. Move **3** spaces to the right. The sum is **6**.

Ten Doubles Facts

You will know ten new addition facts when you learn these doubles:

1 + 1 = 2	6 + 6 = 12
2 + 2 = 4	7 + 7 = 14
3 + 3 = 6	8 + 8 = 16
4 + 4 = 8	9 + 9 = 18
5 + 5 = 10	10 + 10 = 20

= 6

0 + 3 = 3
1 + 3 = 4
2 + 3 = 5
3 + 3 = 6
4 + 3 = 7
5 + 3 = 8
6 + 3 = 9
7 + 3 = 10
8 + 3 = 11
9 + 3 = 12
10 + 3 = 13

10 11 12 13 14 15 16 17 18 19 20

Learn to Add 4

When you add two even numbers, the sum will be even. The number **8** is even. The number **4** is also even.

8 + 4 puppies play fetch.

Begin at **8**. Move **4** spaces to the right. The sum is **12**. The number **12** is even, too.

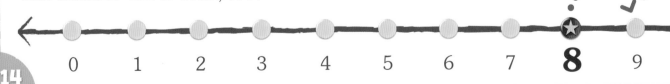

Even Numbers

An even number of puppies comes in pairs. But 0 is an even number, too. Even numbers are 0, 2, 4, 6, 8, 10, 12, 14, 16, 18, 20, and so on.

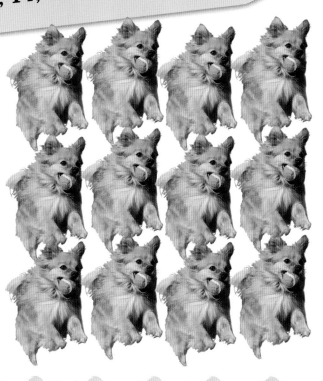

$$= 12$$

$0 + 4 = 4$

$1 + 4 = 5$

$2 + 4 = 6$

$3 + 4 = 7$

$4 + 4 = 8$

$5 + 4 = 9$

$6 + 4 = 10$

$7 + 4 = 11$

$\mathbf{8 + 4 = 12}$

$9 + 4 = 13$

$10 + 4 = 14$

10 11 **12** 13 14 15 16 17 18 19 20

Learn to Add 5

When you add two odd numbers, the sum will always be even. The number **3** is odd. The number **5** is also odd.

3 + 5 kittens scratch the posts.

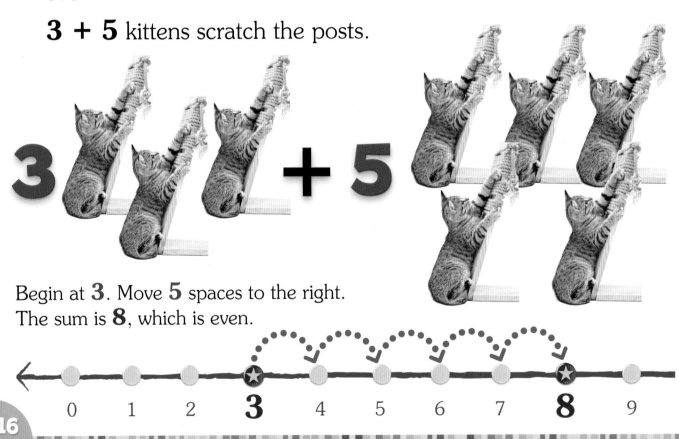

Begin at **3**. Move **5** spaces to the right.
The sum is **8**, which is even.

Odd Numbers

An odd number of kittens doesn't pair up. There will always be one kitten by itself. Odd numbers are 1, 3, 5, 7, 9, 11, 13, 15, 17, 19, and so on.

= 8

$$0 + 5 = 5$$
$$1 + 5 = 6$$
$$2 + 5 = 7$$
$$\mathbf{3 + 5 = 8}$$
$$4 + 5 = 9$$
$$5 + 5 = 10$$
$$6 + 5 = 11$$
$$7 + 5 = 12$$
$$8 + 5 = 13$$
$$9 + 5 = 14$$
$$10 + 5 = 15$$

10 11 12 13 14 15 16 17 18 19 20

Learn to Add 6

When you add an even number and an odd number together, the sum will always be odd. The number **7** is odd. The number **6** is even.

7 + 6 puppies wag their tails.

7 + **6**

Begin at **7**. Move **6** spaces to the right. The sum is **13**, which is odd.

← ● ● ● ● ● ● ● ⭐ ● ●

0   1   2   3   4   5   6   **7**   8   9

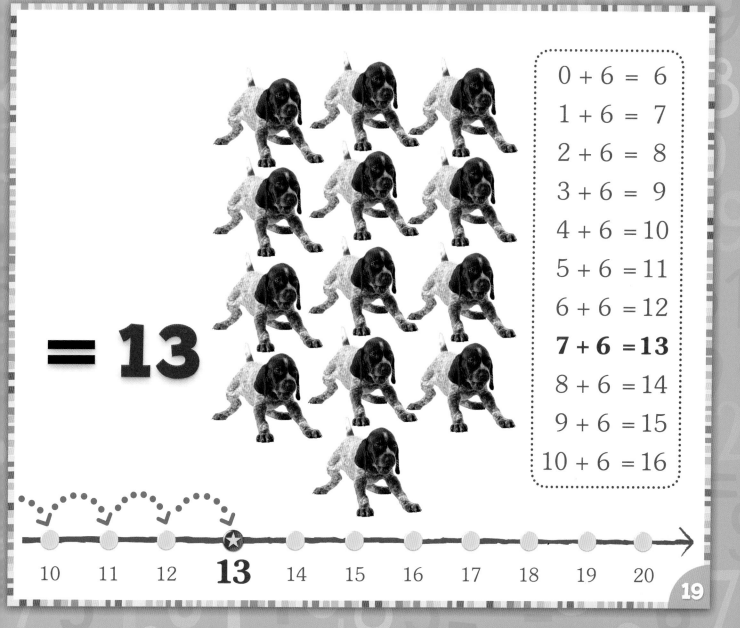

= 13

$$0 + 6 = 6$$
$$1 + 6 = 7$$
$$2 + 6 = 8$$
$$3 + 6 = 9$$
$$4 + 6 = 10$$
$$5 + 6 = 11$$
$$6 + 6 = 12$$
$$\mathbf{7 + 6 = 13}$$
$$8 + 6 = 14$$
$$9 + 6 = 15$$
$$10 + 6 = 16$$

10 11 12 **13** 14 15 16 17 18 19 20

Learn to Add 7

When you learn one addition fact, you actually learn two!
The addition fact **6 + 7** is the same as **7 + 6**.

6 + 7 kittens stretch their toes.

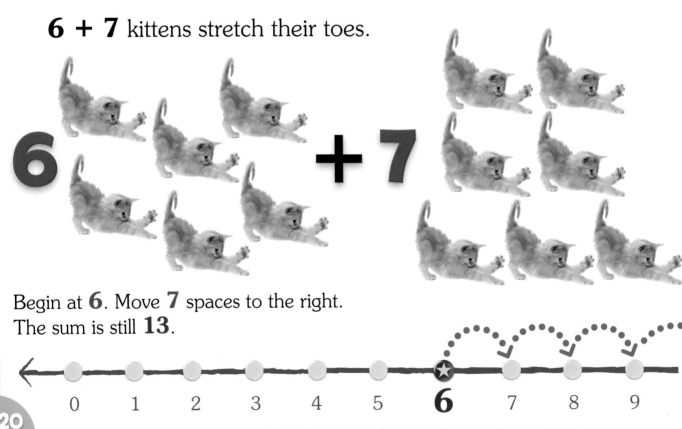

Begin at **6**. Move **7** spaces to the right.
The sum is still **13**.

0 1 2 3 4 5 **6** 7 8 9

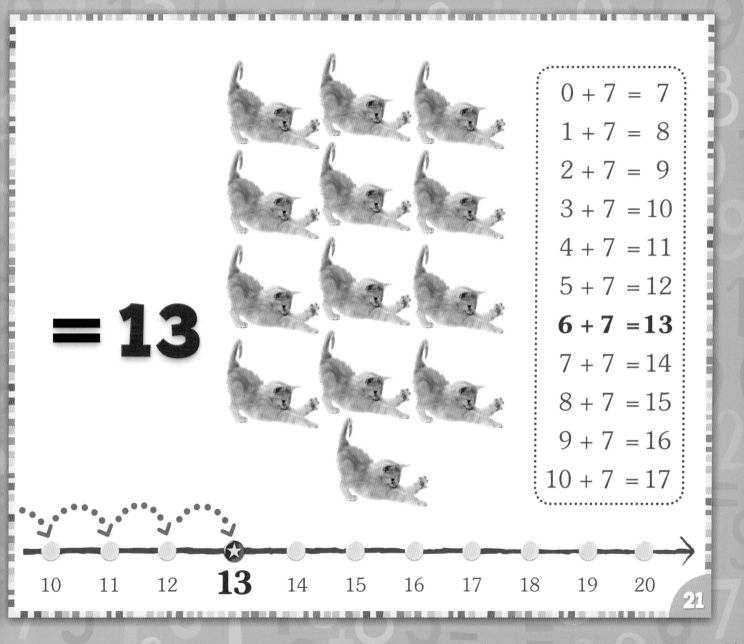

$= 13$

$$0 + 7 = 7$$
$$1 + 7 = 8$$
$$2 + 7 = 9$$
$$3 + 7 = 10$$
$$4 + 7 = 11$$
$$5 + 7 = 12$$
$$\mathbf{6 + 7 = 13}$$
$$7 + 7 = 14$$
$$8 + 7 = 15$$
$$9 + 7 = 16$$
$$10 + 7 = 17$$

10 11 12 **13** 14 15 16 17 18 19 20

Learn to Add 8

The addition fact **1 + 8** has the same sum as **1 + 4 + 4**. This is because **4 + 4 = 8**.

1 + 8 puppies take a bath.

Begin at **1**. Move **8** spaces to the right. The sum is **9**.

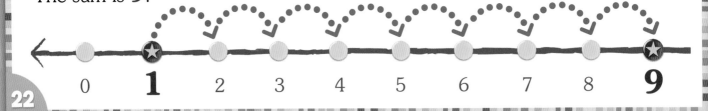

Breaking It Down

If one of your addends is big, you can break it down to two smaller numbers. You already know 4 + 4 = 8, so just do 1 + 4 + 4 to get the same answer as 1 + 8.

= 9

$$0 + 8 = 8$$
$$\mathbf{1 + 8 = 9}$$
$$2 + 8 = 10$$
$$3 + 8 = 11$$
$$4 + 8 = 12$$
$$5 + 8 = 13$$
$$6 + 8 = 14$$
$$7 + 8 = 15$$
$$8 + 8 = 16$$
$$9 + 8 = 17$$
$$10 + 8 = 18$$

10   11   12   13   14   15   16   17   18   19   20

Learn to Add 9

Now that you have learned all about addition, can you solve this addition fact?

6 + 9 kittens meow for you to pet them.

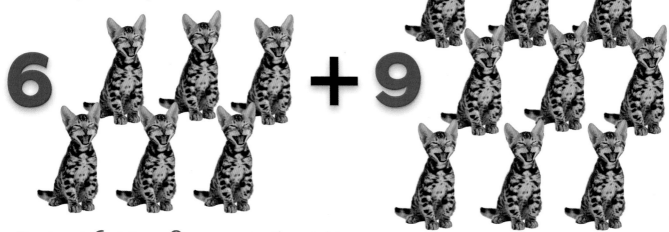

Begin at **6**. Move **9** spaces to the right. The sum is **15**.

0   1   2   3   4   5   **6**   7   8   9

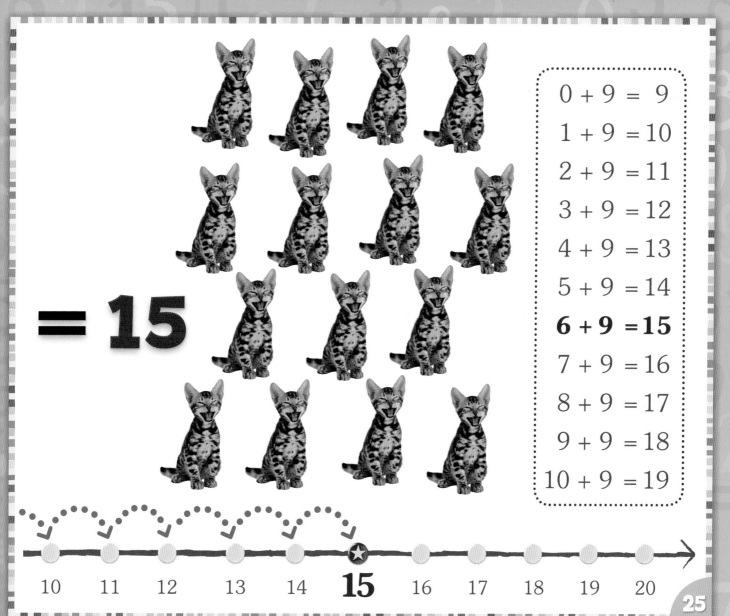

$$= 15$$

$$0 + 9 = 9$$
$$1 + 9 = 10$$
$$2 + 9 = 11$$
$$3 + 9 = 12$$
$$4 + 9 = 13$$
$$5 + 9 = 14$$
$$\mathbf{6 + 9 = 15}$$
$$7 + 9 = 16$$
$$8 + 9 = 17$$
$$9 + 9 = 18$$
$$10 + 9 = 19$$

10   11   12   13   14   **15**   16   17   18   19   20

Learn to Add 10

Try another one! What is the sum of **7 + 10**?

7 + 10 puppies give their paws.

Begin at **7**. Move **10** spaces to the right.
The sum is **17**.

0   1   2   3   4   5   6   **7**   8   9

Making Ten

Can you think of all the facts that add up to 10? There are 11 of them!

$0 + 10 = 10$
$10 + 0 = 10$
$1 + 9 = 10$
$9 + 1 = 10$

$2 + 8 = 10$
$8 + 2 = 10$
$3 + 7 = 10$
$7 + 3 = 10$

$4 + 6 = 10$
$6 + 4 = 10$
$5 + 5 = 10$

$0 + 10 = 10$
$1 + 10 = 11$
$2 + 10 = 12$
$3 + 10 = 13$
$4 + 10 = 14$
$5 + 10 = 15$
$6 + 10 = 16$
$7 + 10 = 17$
$8 + 10 = 18$
$9 + 10 = 19$
$10 + 10 = 20$

$= 17$

10 11 12 13 14 15 16 **17** 18 19 20

Review Addition Facts

Do you know all the addition facts in this book? Take a look!

+ 0	+ 1	+ 2	+ 3	+ 4	+ 5
0 + 0 = 0	0 + 1 = 1	0 + 2 = 2	0 + 3 = 3	0 + 4 = 4	0 + 5 = 5
1 + 0 = 1	1 + 1 = 2	1 + 2 = 3	1 + 3 = 4	1 + 4 = 5	1 + 5 = 6
2 + 0 = 2	2 + 1 = 3	2 + 2 = 4	2 + 3 = 5	2 + 4 = 6	2 + 5 = 7
3 + 0 = 3	3 + 1 = 4	3 + 2 = 5	3 + 3 = 6	3 + 4 = 7	3 + 5 = 8
4 + 0 = 4	4 + 1 = 5	4 + 2 = 6	4 + 3 = 7	4 + 4 = 8	4 + 5 = 9
5 + 0 = 5	5 + 1 = 6	5 + 2 = 7	5 + 3 = 8	5 + 4 = 9	5 + 5 = 10
6 + 0 = 6	6 + 1 = 7	6 + 2 = 8	6 + 3 = 9	6 + 4 = 10	6 + 5 = 11
7 + 0 = 7	7 + 1 = 8	7 + 2 = 9	7 + 3 = 10	7 + 4 = 11	7 + 5 = 12
8 + 0 = 8	8 + 1 = 9	8 + 2 = 10	8 + 3 = 11	8 + 4 = 12	8 + 5 = 13
9 + 0 = 9	9 + 1 = 10	9 + 2 = 11	9 + 3 = 12	9 + 4 = 13	9 + 5 = 14
10 + 0 = 10	10 + 1 = 11	10 + 2 = 12	10 + 3 = 13	10 + 4 = 14	10 + 5 = 15

+ 6	+ 7	+ 8	+ 9	+ 10	Doubles
0 + 6 = 6	0 + 7 = 7	0 + 8 = 8	0 + 9 = 9	0 + 10 = 10	0 + 0 = 0
1 + 6 = 7	1 + 7 = 8	1 + 8 = 9	1 + 9 = 10	1 + 10 = 11	1 + 1 = 2
2 + 6 = 8	2 + 7 = 9	2 + 8 = 10	2 + 9 = 11	2 + 10 = 12	2 + 2 = 4
3 + 6 = 9	3 + 7 = 10	3 + 8 = 11	3 + 9 = 12	3 + 10 = 13	3 + 3 = 6
4 + 6 = 10	4 + 7 = 11	4 + 8 = 12	4 + 9 = 13	4 + 10 = 14	4 + 4 = 8
5 + 6 = 11	5 + 7 = 12	5 + 8 = 13	5 + 9 = 14	5 + 10 = 15	5 + 5 = 10
6 + 6 = 12	6 + 7 = 13	6 + 8 = 14	6 + 9 = 15	6 + 10 = 16	6 + 6 = 12
7 + 6 = 13	7 + 7 = 14	7 + 8 = 15	7 + 9 = 16	7 + 10 = 17	7 + 7 = 14
8 + 6 = 14	8 + 7 = 15	8 + 8 = 16	8 + 9 = 17	8 + 10 = 18	8 + 8 = 16
9 + 6 = 15	9 + 7 = 16	9 + 8 = 17	9 + 9 = 18	9 + 10 = 19	9 + 9 = 18
10 + 6 = 16	10 + 7 = 17	10 + 8 = 18	10 + 9 = 19	10 + 10 = 20	10 + 10 = 20

Activities with Addition

Make Flashcards

Write an addition fact on one side of an index card. Write the answer on the other side. Make as many cards as you want. Test yourself! How many did you get right?

Add Things Around the House

Numbers are everywhere you look. Add the number of shoes to the number of coats you have. Add doors, windows, rooms, or anything you want!

Create a Quiz

List five different addition facts without the sums on a piece of paper. Have a friend solve the problems. Check your friend's answers by looking in this book. Did your friend pass the quiz?

Write a Story

Write your own story using an addition fact. For example:

Billy went to the store with his mom. His mom put 5 apples and 6 bananas in the shopping cart. Billy's mom is buying 11 pieces of fruit all together. She then went to the vegetable part of the store. . . .

You can make it as long or as short as you want. You can even draw pictures to go with your story!

Learn More

Books

May, Eleanor. **Albert Doubles the Fun: Adding Doubles.** New York, NY: Kane Press, 2017.

Steffora. Tracey. **Adding with Ants.** Oxford, UK: Raintree, 2013.

Summers, Portia. **Adding Coins and Bills.** New York, NY: Enslow Publishing, 2016.

Websites

ABCYa.com
www.abcya.com/first_grade_computers.htm#numbers-cat
Play games to practice math!

Doctor Genius, Addition
www.mathabc.com/math-1st-grade/addition/addition-up-to-20
Solve addition problems online!

Math Playground
www.mathplayground.com/grade_1_games.html
Here are more fun math games!

Index

Published in 2018 by Enslow Publishing, LLC.
101 W. 23rd Street, Suite 240, New York, NY 10011

Copyright © 2018 by Enslow Publishing, Inc.

Library of Congress Cataloging-in-Publication Data

Names: Moldovo, Eustacia, author. | Murphy, Patricia J., 1963 – author.
Title: Learning addition with puppies and kittens / Eustacia Moldovo, Patricia J. Murphy.
Description: New York, NY : Enslow Publishing, 2018. | Series: Math fun with puppies and kittens | Audience: K to grade 3. | Includes bibliographical references and index.
Identifiers: LCCN 2017012587 | ISBN 9780766090873 (library bound) | ISBN 9780766090705 (pbk.) | ISBN 9780766090699 (6 pack)
Subjects: LCSH: Addition—Juvenile literature. | Mathematics—Juvenile literature.
Classification: LCC QA115 .M645 2018 | DDC 513.2/11—dc23
LC record available at https://lccn.loc.gov/2017012587

Printed in China

To Our Readers: We have done our best to make sure all websites in this book were active and appropriate when we went to press. However, the author and the publisher have no control over and assume no liability for the material available on those websites or on other websites they may link to. Any comments or suggestions can be sent by email to customerservice@enslow.com or to the address on the back cover.

Portions of this book appeared in the book *Adding Puppies and Kittens*.

Photo credits: Cover, p. 1 (puppy) Eric Isselee/Shutterstock.com; cover, p. 1 (kittens) Dmitry Kalinovsky/Shutterstock.com; pp. 2 (kitten, far left), 27 (kitten) Happy monkey/Shutterstock.com; pp. 2 (puppy, left), 29 Liliya Kulianionak/Shutterstock.com; p. 2 (other animals) Ermolaev Alexander/Shutterstock.com; p. 4 (puppies) Jagodka/Shutterstock.com; p. 4 (kittens) Gladkova Svetlana/Shutterstock.com; p. 5 Tatiana Katsa/Shutterstock.com; pp. 6, 7 (puppies) Brberrys/Shutterstock.com; p. 7 (kitten) Lubava/Shutterstock.com; pp. 8, 9 Tsekhmister/Shutterstock.com; pp. 10, 11 Dora Zett/Shutterstock.com; pp. 12, 13 (kitten) absolutimages/Shutterstock.com; p. 13 (puppy) Artsilense/Shutterstock.com; p. 14, 15 (puppy) gwb/Shutterstock.com; p. 15 (kitten) yevgeniy11/Shutterstock.com; pp. 16, 17 (kitten) Sonsedska Yuliia/Shutterstock.com; p. 17 (puppy) Kalamurzing/Shutterstock.com; pp. 18, 19, 26, 27 (puppy) WilleeCole Photography/Shutterstock.com; pp. 20, 21 cynoclub/Shutterstock.com; pp. 22, 23 (puppy) Africa Studio/Shutterstock.com; p. 23 (kitten) Oksana Kuzmina/Shutterstock.com; pp. 24, 25 Adya/Shutterstock.com.